WORKBOOK

Working for over 25 YEARS WITH Cambridge Assessment International Education

Cambridge Assessment
International Education

Endorsed for learner support

Cambridge IGCSE™

Mathematics
Core and Extended

Fourth edition

Ric Pimentel
Terry Wall

T0272715

HODDER
EDUCATION

Introduction

Welcome to the Cambridge IGCSE™ Mathematics Core and Extended Workbook. The aim of this Workbook is to provide you with further opportunity to practise the skills you have acquired while using the Cambridge IGCSE Mathematics Core and Extended Student Book. It is designed to complement the fourth edition of the Student Book and to provide additional exercises to help you in your preparation for Cambridge IGCSE Mathematics examinations.

The topics in this Workbook reflect the topics in the Student Book, and the exercises follow the same numbering as the Student Book. There is no set way to approach using this Workbook. You may wish to use it to supplement your understanding of the different topics as you work through each topic in the Student Book, or you may prefer to use it to reinforce your skills in dealing with particular topics as you prepare for your examination. The Workbook is intended to be sufficiently flexible to suit whatever you feel is the best approach for your needs.

The sections of the Workbook relating to the Extended content have been marked with an asterisk sign * next to the exercise or question number. If the entire chapter deals only with Extended content, you will also see the asterisk sign * next to the chapter title.

The questions, example answers, marks awarded and/or comments that appear in this book were written by the authors. In examination, the way marks would be awarded to answers like these may be different.

Every effort has been made to trace all copyright holders, but if any have been inadvertently overlooked, the Publishers will be pleased to make the necessary arrangements at the first opportunity.

Although every effort has been made to ensure that website addresses are correct at time of going to press, Hodder Education cannot be held responsible for the content of any website mentioned in this book. It is sometimes possible to find a relocated web page by typing in the address of the home page for a website in the URL window of your browser.

Hachette UK's policy is to use papers that are natural, renewable and recyclable products and made from wood grown in well-managed forests and other controlled sources. The logging and manufacturing processes are expected to conform to the environmental regulations of the country of origin.

Orders: please contact Hachette UK Distribution, Hely Hutchinson Centre, Milton Road, Didcot, Oxfordshire, OX11 7HH. Telephone: +44 (0)1235 827827. Email education@hachette.co.uk Lines are open from 9 a.m. to 5 p.m., Monday to Friday. You can also order through our website: www.hoddereducation.com

© Ric Pimentel and Terry Wall 2013, 2018

First published in 2013
This edition published in 2018 by
Hodder Education
An Hachette UK Company
Carmelite House
50 Victoria Embankment
London EC4Y 0DZ

Impression number   10 9 8 7 6

Year   2022

All rights reserved. Apart from any use permitted under UK copyright law, no part of this publication may be reproduced or transmitted in any form or by any means, electronic or mechanical, including photocopying and recording, or held within any information storage and retrieval system, without permission in writing from the publisher or under licence from the Copyright Licensing Agency Limited. Further details of such licences (for reprographic reproduction) may be obtained from the Copyright Licensing Agency Limited, www.cla.co.uk

Cover photo © Shutterstock/ju.grozyan

Illustrations by Datapage (India) Pvt. Ltd. and Integra Software Services Pvt. Ltd.

Typeset by Integra Software Services Pvt. Ltd., Pondicherry, India

Printed in the UK

A catalogue record for this title is available from the British Library.

ISBN: 978 1 5104 2170 7

Contents

1 Number and language

Exercises 1.1–1.5

1 List the prime factors of these numbers and express each number as a product of its prime factors.

a 25 ....[2]

b 48 ....[2]

2 Write the reciprocal of the following.

a $\frac{2}{7}$ ....[1]   **b** $\frac{13}{5}$ ....[1]

Exercise 1.6

1 State whether each of the following is a rational or irrational number.

a $\sqrt{5} \times \sqrt{2}$ ....[1]   **b** $\frac{\sqrt{16}}{4}$ ....[1]

Exercises 1.7–1.10

1 Without a calculator, work out the following.

a $\sqrt{5\frac{4}{9}}$

b $\sqrt[3]{15\frac{5}{8}}$

....[2]   ....[2]

Exercise 1.11

1 A plane flying at 8500 m drops a sonar device onto the ocean floor. If the sonar falls a total of 10 200 m, how deep is the ocean at this point?

....[2]

 Photocopying prohibited   *Cambridge IGCSE™ Core and Extended Mathematics Workbook*

2 Accuracy

Exercises 2.1–2.3

1 Write the following to the number of significant figures written in brackets.

a 15.01 (1 s.f.)

b 0.04299 (2 s.f.)

c 3.04901 (3 s.f.)

.... [1]

.... [1]

.... [1]

Exercise 2.5

1 A car holds 70 litres of petrol correct to the nearest litre. Its fuel economy is 12 km per litre to the nearest kilometre. Write down, but do not work out, the calculation for the upper limit of the distance the car can travel.

.... [2]

Exercise 2.6*

1 Calculate upper and lower bounds for the following calculations; each of the numbers is given correct to the nearest whole number.

a 15×25

b 128×22

.... [2]

.... [2]

c 1000×5

d $\frac{3}{4}$

.... [2]

.... [2]

e $\frac{120}{60}$

.... [2]

2 Calculate upper and lower bounds for the following calculations; each of the numbers is given correct to one decimal place.

a $2.4 + 14.1$

b 3.3×8.8

.... [2]

.... [2]

c 100.0×4.9

d $(0.4 - 0.1)^2$

.... [2]

.... [2]

3 If $a = 18$ and $b = 22$, both correct to the nearest whole number, between what limits is $\sqrt{a^2 + b^2}$?

.... [4]

Exercise 2.7*

1 A town is built on a rectangular plot of land measuring 3.7 km by 5.2 km, correct to 1 d.p. What are the upper and lower limits for the area of the town?

.... [2]

2 303 degrees Kelvin is equivalent to 30 degrees Celsius. Both figures are correct to the nearest degree.

a What is the maximum percentage error in each case?

.... [2]

b Explain why the percentage errors are different.

.... [1]

3 1 mile equals 1.6093 km, correct to 4 d.p. If a distance is 7 miles correct to the nearest mile, between what limits is the distance in kilometres?

.... [4]

3 Calculations and order

Exercises 3.1–3.2

1 Write these decimals in order of magnitude, starting with the smallest.

0.055   5.005   5.500   0.505   0.550

.... [1]

Exercises 3.3–3.5

1 In the calculation below, insert any brackets that are needed to make it correct. Check your answer with a calculator.

15   ÷   3   +   2   ÷   2   =   1.5   [3]

4 Integers, fractions, decimals and percentages

Exercises 4.1–4.4

1 Without a calculator, write the following fractions as decimals.

   a $3\frac{9}{20}$ ....[1]

   b $7\frac{19}{25}$ ....[1]

Exercises 4.6–4.10

1 Without a calculator, work out this calculation. Give your answer as a fraction in its simplest form.

   $\left(\frac{4}{9} - 1\frac{4}{5}\right) \div \frac{2}{3}$

   ....[3]

Exercise 4.11*

1 Convert each of the following recurring decimals to fractions in their simplest form:

   a $0.5\dot{6}$ b $1.30\dot{8}$

   .... [2]   .... [3]

2 Without a calculator, evaluate $0.\dot{3}\dot{8} - 0.2\dot{5}$. Convert each decimal to a fraction first.

   ....[4]

5 Further percentages

Exercises 5.1–5.3

1 Express the fraction $\frac{7}{8}$ as a percentage. ....[1]

2 Work out 62.5% of 56. ....[2]

3 Petrol costs 78.5 cents/litre, and 61 cents of this is tax. Calculate the percentage of the cost that is tax.

....[2]

Exercise 5.5*

1 Calculate the value of X in each of the following.

a 65% of X is $292.50

.... [1]

b 15% of X is $93.00

.... [1]

c X% of 12 is 40.8.

.... [1]

d X% of 20 is 32

.... [1]

2 In a school, 45% of the students are boys. If there are 117 boys, calculate the total number of students.

....[2]

3 In an exam, Paulo scored 68%. If he got 153 marks in total, calculate the number of marks available in the exam.

....[2]

4 An elastic band can increase its natural length by 625% when fully stretched. If the fully stretched elastic band has a length of 29 cm, calculate its natural length.

....[3]

6 Ratio and proportion

Exercises 6.2–6.4

1 4g of copper is mixed with 5g of tin.

   a What fraction of the mixture is tin? .... [1]

   b How much tin is there in 1.8 kg of the same mixture? .... [1]

2 The angles of a hexagon add up to 720° and are in the ratio 1:2:4:4:3:1.

   Find the sizes of the largest and smallest angles.

   .... [2]

3 A train travelling at 160 km/h takes 5 hours for a journey. How long would it take a train travelling at 200 km/h?

   .... [2]

Exercise 6.5*

1 Increase 250 by the following ratios.

   a 8:5   **b** 12.5:5

   .... [1]   .... [1]

2 Increase 75 by the following ratios.

   a 7.5:3   **b** 5:2

   .... [1]   .... [1]

3 Decrease 120 by the following ratios.

   a 2:3   **b** 1:4

   .... [1]   .... [1]

Exercise 6.6*

1 A photograph measuring 12 cm by 8 cm is enlarged by a ratio of 9:4. What are the dimensions of the new print?

   .... [2]

 Photocopying prohibited   *Cambridge IGCSE™ Core and Extended Mathematics Workbook*

2 A rectangle measuring 24 cm by 12 cm is enlarged by a ratio of 3 : 2.

    a What is the area of:

        i the original rectangle   **ii** the enlarged rectangle?

        ....[1]   .... [2]

    b By what ratio has the area been enlarged?

        .... [1]

3 A cuboid measuring 12.5 cm by 5 cm by 2.5 cm is enlarged by a ratio of 4 : 1.

    a What is the volume of:

        i the original cuboid   **ii** the enlarged cuboid?

        .... [1]   .... [2]

    b By what ratio has the volume been increased?

        .... [1]

Exercise 6.7

1 In 2017, the world record for the 100 m sprint stood at 9.58 seconds. Calculate the average speed of the record holder:

    a in m/s   **b** in km/h.

    .... [1]   .... [2]

2 The country with the lowest population density in the world is Greenland. Greenland has an area of 2 166 000 km^2 and a population of 55 984. Calculate its population density correct to 2 s.f.

    .... [2]

3 A cube of copper has a mass of 600 kg. Copper has a density of 9.3 g/cm^3. Calculate the length of each side of the cube to the nearest mm.

    .... [3]

7 Indices and standard form

Exercises 7.1–7.4

1 Simplify the following using indices.

$2 \times 7 \times 7 \times 7 \times 7 \times 11 \times 11$ .... [1]

2 Simplify the following.

$16^3 \times 16^{-2} \times 16^{-2}$ .... [2]

3 Work out the following without a calculator.

3×10^{-2} .... [2]

4 Work out the following without a calculator.

144×6^{-2} .... [2]

5 Find the value of x.

$10^x = 1\,000\,000$ .... [2]

6 Find the value of z in each of the following.

  a $3^{(z+2)} = 81$ **b** $2^{-z} = 128^{-1}$

  .... [2]   .... [2]

Exercises 7.5–7.6

1 Write 463 million in standard form. .... [1]

2 Write 0.000 000 000 367 in standard form. .... [1]

3 Deduce the value of x.

$0.04^x = 1.024 \times 10^{-7}$ .... [2]

Exercise 7.7*

Evaluate the following without using a calculator.

1 $49^{\frac{1}{2}}$ **2** $225^{\frac{1}{2}}$

  .... [1]   .... [1]

3 $125^{\frac{1}{3}}$ **4** $1\,000\,000^{\frac{1}{3}}$

  .... [1]   .... [1]

 Photocopying prohibited   *Cambridge IGCSE™ Core and Extended Mathematics Workbook*

5 $343^{\frac{1}{3}}$

.... [2]

6 $625^{\frac{1}{4}}$

.... [2]

7 $81^{\frac{1}{4}}$

.... [2]

8 $1728^{\frac{1}{3}}$

.... [2]

Exercise 7.8*

Work out the following without using a calculator.

1 $\dfrac{17^0}{2^2}$

.... [2]

2 $\dfrac{27^{\frac{2}{3}}}{3^2}$

.... [3]

3 $\dfrac{64^{\frac{1}{2}}}{4^2}$

.... [2]

4 $\dfrac{1^0}{2^3}$

.... [2]

5 $\dfrac{4^{\frac{1}{2}}}{2^2}$

.... [2]

6 $64^{-\frac{1}{2}} \times 2^3$

.... [3]

7 $121^{-\frac{1}{2}} \times 11^2$

.... [3]

8 $729^{-\frac{1}{3}} \div 3^{-2}$

.... [3]

9 $4^{\frac{1}{2}} \times 4^{-2} \times \frac{1}{4}$

.... [3]

10 $27^{\frac{1}{3}} \times 81^{-2}$

.... [3]

8 Money and finance

Exercise 8.1

1 The exchange rate for €1 into Sri Lankan rupees is €1 to 162 rupees. Convert 1000 Sri Lankan rupees to euros.

.... [1]

Exercises 8.2–8.4

1 A caravan is priced at $9500. The supplier offers customers two options for buying it:

Option 1: A 25% deposit followed by 24 monthly payments of $350

Option 2: 36 monthly payments of $380.

a Calculate the amount extra, compared with the cash price, a customer would have to pay with each option.

.... [3]

b Explain why a customer might choose the more expensive option.

....

.... [2]

Exercise 8.6*

1 A couple borrow $140000 to buy a house at 5% compound interest for three years. How much will they owe at the end of the three years?

.... [3]

2 A boat has halved in value in three years. What was the percentage loss in compound terms?

.... [3]

 Photocopying prohibited   *Cambridge IGCSE™ Core and Extended Mathematics Workbook*

9 Time

Exercise 9.1

1 A plane travels 7050 km at an average speed of 940 km/h. If it lands at 13 21, calculate the time it departed.

.... [3]

10 Set notation and Venn diagrams

Exercise 10.1

1 For the set {Moscow, London, Cairo, New Delhi, …}:

  a Describe this set in words:

  .... [1]

  b Write down two more elements of this set.

  .... [2]

Exercise 10.2*

1 The set $A = \{x: 0 < x < 10\}$.

  a List the subset B {prime numbers} .... [2]

  b List the subset C {square numbers} .... [2]

2 $P = \{a, b, c\}$

  a List all the subsets of P. .... [2]

  b List all the proper subsets of P. .... [1]

Exercise 10.3*

1 If $\mathscr{E}$ = {girls' names} and M = {girls' names beginning with the letter A}, what is the set represented by M'?

  .... [1]

2 The Venn diagram below shows the relationship between three sets of numbers A, B and C.

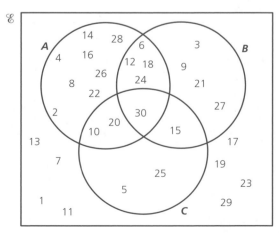

a If $\mathscr{E} = \{1, 2, 3, 4, \ldots, 30\}$, complete the following statements.

  i $A =$ .... [1]

  ii $B =$ .... [1]

  iii $C =$ .... [1]

b Complete the following by entering the correct numbers:

  i $A \cap B = \{$ ....$\}$ [1]

  ii $B \cup C = \{$ ....$\}$ [1]

  iii $A \cap B \cap C = \{$ ....$\}$ [1]

  iv $A' \cap C = \{$ ....$\}$ [2]

3 Consider the Venn diagram below showing the relationship between the sets W, X, Y and Z.

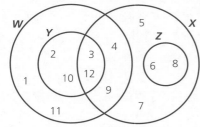

a Complete the following by entering the correct numbers:

  i $X \cup Y = \{$ ....$\}$ [1]

  ii $W \cap X = \{$ ....$\}$ [1]

b Which of the named sets is a subset of X? .... [1]

Exercise 10.4

1 The sets given below represent the letters of the alphabet in each of three English cities.

$P = \{c,a,m,b,r,i,d,g,e\}$, $Q = \{b,r,i,g,h,t,o,n\}$ and $R = \{d,u,r,h,a,m\}$

a Draw a Venn diagram to illustrate this information.   [3]

b Complete the following statements:

i $Q \cup R = \{$ ....$\}$ [1]

ii $P \cap Q \cap R = \{$ ....$\}$ [1]

2 In a class of 30 students, 16 do athletics (A), 17 do swimming (S), whilst 3 do neither.

a Complete the Venn diagram below.

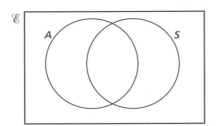

[3]

b Calculate the following:

i n($A \cap S$) .... [1]   **ii** n($A \cup S$)' .... [1]

Exercise 10.5*

1 A class of 15 students was asked what pets they had. Each student had either a bird (B), cat (C), fish (F), or a combination of them.

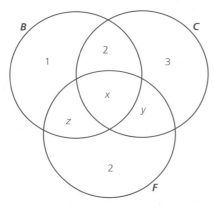

a If n(B) = 10, n(C) = 11 and n(F) = 9, calculate:

i z

.... [3]

ii x

.... [2]

iii y

.... [1]

b Calculate n($C \cup F$) .... [1]

11 Algebraic representation and manipulation

Exercises 11.1–11.3

1 Expand the following and simplify where possible.

    a $-5(x+4)$ .... [1]

    b $\frac{1}{2}(8x+4)+2(3x+6)$ .... [2]

2 Expand and simplify the following.

    a $3q(2+7r)+2r(3+4q)$ .... [2]

    b $(a+8)(a+4)$ .... [2]

    c $(j+k)(k-m)$ .... [2]

Exercise 11.4

Factorise the following.

1 $42x^2-21xy^2$ .... [2]

2 $m^3-m^2n-n^2m$ .... [2]

Exercise 11.5

1 Evaluate the expressions below if: $p=3, q=-3, r=-1$ and $s=5$.

    a $p^2+q^2+r^2+s$

    .... [2]

    b $-p^3-q^3-r^3-s^3$

    .... [2]

Exercise 11.6

1 Make the letter in bold the subject of the formula.

    a $a\mathbf{b}-c=d$ **b** $\dfrac{p}{-\mathbf{q}}+r=-s$

    .... [2]   .... [2]

Exercise 11.7*

Expand and simplify:

1 $(2d + 3)(2d - 3)$ .... [2]

2 $(3e - 7)(3e - 7)$ .... [2]

3 $(2f + 3g)(2f - 3g)$ .... [2]

4 $(4 - 5h)(5h + 4)$ .... [2]

5 $(2x + 1)(3x - 1)$ .... [2]

6 $(x - 4)(x + 2)(3x - 2)$

.... [2]

7 $(2x^2 - 3x + 1)(4x - 1)$

.... [2]

Exercise 11.8*

Factorise by grouping.

1 $ac + a + b + bc$

.... [2]

2 $3cd + 3d + 4e + 4ce$

.... [2]

3 $fg - 4f - 6g + 24$

.... [2]

4 $p^2 - 2pq - 2pr + 4rq$

.... [2]

5 $16m^2 + 44mn + 121n + 44m$

.... [2]

Exercise 11.9*

Factorise:

1 $16m^2 - 121n^2$

.... [2]

2 $x^6 - y^6$

.... [2]

3 $9a^4 - 144b^4$

.... [2]

4 $81m^2 - 16n^2$

.... [2]

 Photocopying prohibited *Cambridge IGCSE™ Core and Extended Mathematics Workbook*

Exercise 11.10*

Evaluate by factorising.

1 $17^2 - 16^2$ .... [2]

2 $3^4 - 1$ .... [2]

3 $98^2 - 4$ .... [2]

Exercise 11.11*

1 Factorise the following quadratic expressions.

a $a^2 + 5a + 6$

b $b^2 - 3b - 10$

.... [2]

.... [2]

c $c^2 - 10c + 16$

d $d^2 - 18d + 81$

.... [2]

.... [2]

2 Factorise the following quadratic expressions.

a $2e^2 + 3e + 1$

b $3f^2 + f - 2$

.... [3]

.... [3]

c $2g^2 - g - 1$

d $9h^2 - 4$

.... [3]

.... [3]

e $j^2 + 4jk + 4k^2$

.... [3]

Exercises 11.12–11.13*

In the following formulas, make 'a' the subject.

1 $\dfrac{p}{q} = \dfrac{2xa}{r}$

2 $\dfrac{ma^2}{3n} = \dfrac{2}{n}$

.... [3]

.... [3]

3 $t = \dfrac{2r\sqrt{a}}{b}$

4 $t = \dfrac{2p\sqrt{b}}{a}$

.... [3]   .... [3]

5 $\dfrac{2\sqrt{a}}{3} = \dfrac{b^2}{c}$

.... [3]

Exercise 11.14*

1 The circumference of a circle is given by the equation $C = 2\pi r$. Find r when C is 18.8 cm.

.... [4]

2 The area of a circle is given by the equation $A = \pi r^2$. Find r when $A = 78.5\,\text{cm}^2$.

.... [4]

3 A parallelogram has area $48\,\text{cm}^2$ and one side 8 cm. Find the perpendicular height of the parallelogram using the formula $A = lp$, where l is the length of a side and p is the perpendicular height.

.... [4]

4 The surface area of a cylinder is $188\,\text{cm}^2$. Its radius is 3 cm. Given the formula $A = 2\pi r(r + h)$, rearrange to find an expression for h, then find the value of h. Draw a sketch if necessary.

.... [4]

5 A cylinder of radius 4 cm has a volume of $503\,\text{cm}^3$. Given the formula $V = \pi r^2 h$, rearrange to find an expression for h, then find the value of h. Draw a sketch if necessary.

.... [4]

Exercise 11.15*

Simplify the following fractions.

1 $\dfrac{2a^2}{3} \times \dfrac{6}{a}$

2 $\dfrac{5c^2}{2d} \times \dfrac{6e}{c} \times \dfrac{d^2}{c}$

.... [2]   .... [2]

3 $\dfrac{9p}{7} \times \dfrac{14}{3p}$

4 $\dfrac{4x^3}{3y^4} \times \dfrac{6y^5}{2x^2}$

.... [2]   .... [2]

Exercises 11.16–11.17*

Simplify the following fractions.

1 $\dfrac{a}{4} + \dfrac{b}{3}$

2 $\dfrac{3c}{4} - \dfrac{2c}{3}$

.... [2]   .... [2]

3 $\dfrac{f}{9} + f$

4 $\dfrac{3e}{7} - \dfrac{2e}{3}$

.... [2]   .... [2]

Exercise 11.18*

Simplify the following fractions.

1 $\dfrac{1}{p+3} + \dfrac{2}{p-1}$

2 $\dfrac{a(a+5)}{b(a+5)}$

.... [3]   .... [3]

3 $\dfrac{a^2 - 3a}{(a+1)(a-3)}$

4 $\dfrac{a^2 + 2a}{a^2 + 5a + 6}$

.... [3]   .... [3]

5 $\dfrac{a^3 - a}{a^2 - 1}$

.... [3]

12 Algebraic indices

Exercises 12.1–12.2

1 Simplify $p^4 \times q^7 \times p^3 \times q^2 \times r$ using indices.

.... [2]

2 Simplify.

a $ac^5 \times ac^3$

.... [2]

b $3(2b^3)^3$

.... [2]

Exercise 12.3*

1 Rewrite the following in the form $a^{\frac{m}{n}}$.

a $\left(\sqrt[4]{a}\right)^5$ .... [2]

b $\left(\sqrt{a}\right)^7$ .... [2]

2 Rewrite the following in the form $\left(\sqrt[n]{b}\right)^m$.

a $b^{-\frac{3}{5}}$ .... [2]

b $b^{\frac{7}{9}}$ .... [2]

3 Simplify the algebraic expressions, giving your answer in the form $a^{\frac{m}{n}}$.

a $a^{\frac{2}{3}} \times a^{-\frac{3}{4}}$ .... [2]

b $\dfrac{a^{-3}}{\sqrt[3]{a}}$ .... [2]

c $\dfrac{\left(a^{-2}\right)^3}{a^{-\frac{5}{2}} \times \left(\sqrt[3]{a}\right)^{-2}}$ .... [3]

Photocopying prohibited   *Cambridge IGCSE™ Core and Extended Mathematics Workbook*

13 Equations and inequalities

Exercise 13.1

1 Solve the following linear equations.

a $3c - 9 = 5c + 13$

b $\dfrac{4(h + 5)}{3} = 12$

.... [2]   .... [2]

2 Solve this linear equation.

$$\dfrac{7 - 2j}{5} = \dfrac{11 - 3j}{8}$$

.... [2]

Exercise 13.2

The interior angles of a regular pentagon sum to 540°.

1 A pentagon has angles $(4x + 20)°$, $(x + 40)°$, $(3x - 50)°$, $(3x - 130)°$ and 110° as shown.

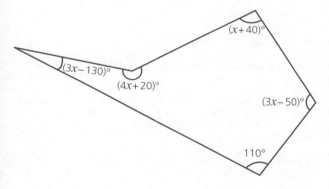

Find the value of each angle.

.... [3]

Exercises 13.3–13.5

1 Solve the simultaneous equations:

a $-5p - 3q = -24$
  $-5p + 3q = -6$

.... [3]

b $2a + 3b = 12$
  $a + b = 5$

c $4h + j = 14$
  $12h - 6j = 6$

.... [3]   .... [4]

2 If a girl multiplies her age in years by 4 and adds three times her brother's age, she gets 64. If the boy adds his age in years to double his sister's age, he gets 28. How old are they?

.... [4]

3 A square has sides $2x$, $40 - 3x$, $25 + 3y$ and $10 - 2y$. Calculate the area of the square.

.... [3]

Exercise 13.6

1 Zach is two years older than his sister Leda and three years younger than his cat, Smudge.

a Where Zach's age is x, write expressions for the ages of Leda and Smudge in terms of x.

.... [4]

b Find their ages if their total age is 22 years.

.... [4]

2 A number squared has the number squared then doubled added to it. The total is 300.

Find two possible values for the number.

.... [4]

Exercise 13.7*

1 Solve the following equations and give two solutions for x.

a $x^2 + x - 12 = 0$

b $x^2 - 9x + 18 = 0$

.... [3]   .... [3]

c $x^2 + 10x + 21 = 0$

d $x^2 = -(3x + 2)$

.... [3]   .... [3]

e $x^2 - 2x = 35$

f $-42 + 13x = x^2$

.... [3]

.... [3]

g $x^2 - 169 = 0$

h $x^2 - 40 = 9$

.... [3]

.... [3]

Exercise 13.8*

Solve the following quadratic equations where possible.

1 $2x^2 + 8x + 6 = 0$

2 $3x^2 + 4x = -1$

.... [4]

.... [4]

3 $5x^2 = 4x + 1$

4 $3x^2 = 108$

.... [3]

.... [3]

5 $3x^2 = 27$

6 $3x^2 = -36$

.... [3]

.... [3]

7 $3x^2 = -108$

8 $4x^2 = 1$

.... [3]

.... [3]

9 $16x^2 = 1$

10 $3x^2 = \dfrac{4}{3}$

.... [3]

.... [3]

11 $25x^2 = 64$

12 $16x^2 = -64$

.... [3]   .... [3]

Exercise 13.9*

1 I have a number of dollars in my pocket. If I square the amount, it is the same as 21 dollars more than four times the amount. How much do I have?

.... [4]

2 A triangle has base length $2b$ cm and a height 2 less than $2b$ cm. Its area is 60 cm^2 as shown. What is the base length and height?

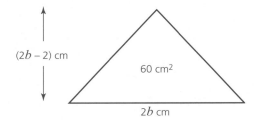

$(2b - 2)$ cm

60 cm^2

$2b$ cm

.... [4]

3 A right-angled triangle has a hypotenuse of 13 cm and two shorter sides of x cm and $x + 7$ cm. Use Pythagoras' theorem to find the length of the two sides.

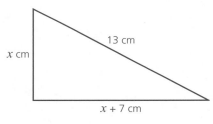

13 cm

x cm

$x + 7$ cm

.... [4]

4 Two consecutive numbers multiply to make 552. Find the numbers.

.... [4]

5 A man buys a number of golf balls for $6. If he had paid 50 cents less for each, he could have bought six more for $6. How many balls did he buy?

.... [4]

Exercise 13.10*

Solve the quadratic equations either by using the quadratic formula.

1 $6x^2 + 22x = -12$

2 $x^2 = -(1 + 4x)$

.... [4]   .... [4]

3 $10x - 2 = -4x^2$

4 $3 - x = 3x^2$

.... [4]   .... [4]

5 $15 = 7x + 2x^2$

6 $4y^2 = -5y - 1$

.... [4]   .... [4]

7 $8x - 4x^2 = -6$

8 $10x^2 = 60 - 25x$

.... [4]   .... [4]

9 $2x^2 + 6.6x - 1.4 = 0$

10 $2x(x + 1) = x^2 - 2x - 4$

.... [4]   .... [4]

Exercise 13.11*

In questions 1–3, solve the equations simultaneously.

1 $y = 6x - 10$
$y = x^2 - 4x + 11$

2 $y = -3$
$y = (x + 4)^2 - 3$

3 $y = 2x - 7$
$y = \dfrac{4}{x}$

.... [4]   .... [4]   .... [4]

4 The area of triangle A and the perimeter of the equilateral triangle B below have the same numerical value. Calculate the value(s) of x.

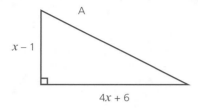 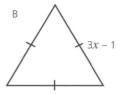

.... [4]

Exercises 13.12–13.13*

Write the following as linear inequalities using the correct mathematical symbols and show the solution on a number line.

1 16 plus $2x$ is less than 10

.... [4]

2 19 is greater than or equal to $9x$ plus 1

.... [3]

3 1 minus $3x$ is equal to or exceeds 13

.... [3]

4 A half x is smaller than 2

.... [3]

5 A third of x is equal to or bigger than 1

.... [3]

6 $4x$ is more than 8 but less than 16

.... [4]

7 $9x$ is between 9 and 45 but not equal to either

.... [4]

8 $2x - 6$ is between 4 and 10 but equal to neither

.... [4]

9 3 is equal to or is less than $2x + 1$, which is less than 9

.... [4]

10 20 is equal to or bigger than $2x - 5$, which is greater than or equal to 10

.... [4]

14 Linear programming*

Exercise 14.1*

Solve the inequalities:

1 $4x - 12 \geqslant -6$ **2** $2 \leqslant -3x + 8$ **3** $9 < -3(y + 4) \leqslant 15$

..................................[1]   ....[1]   ....[2]

Exercise 14.2*

In each question, shade the region which satisfies the inequality on the axes given.

1 $y \leqslant 2$

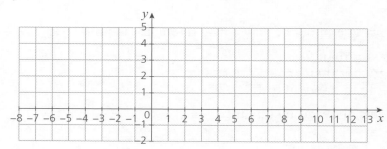

[1]

2 $y + 2x - 4 < 0$

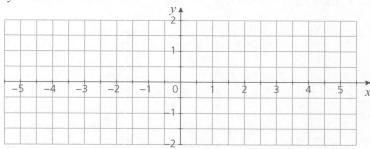

[2]

3 $x - 3y \geqslant 6$

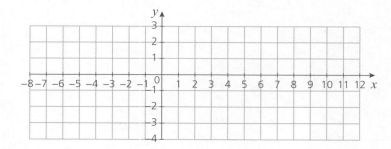

[3]

Exercise 14.3*

On the same pair of axes, plot the following inequalities and leave unshaded the region which satisfies all of them simultaneously.

1 $y \leqslant -2x - 3$, $y > -\frac{1}{2}x - 3$, $x \geqslant -3$

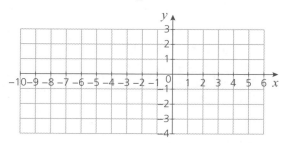

[3]

2 $y \leqslant -\frac{1}{2}x + 2$, $y \geqslant 0$, $3y + 2x - 6 > 0$

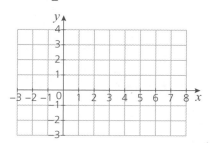

[3]

Exercise 14.4*

1 A team at the Olympics has the following number of male (x) and female (y) athletes:

- the number of male athletes is greater than 5
- the number of female athletes is greater than 7
- the total number of athletes is less than or equal to 15.

a Express each of the three statements above as inequalities.

.... [3]

b On the axes below, identify the region that satisfies all the inequalities by shading the unwanted regions.

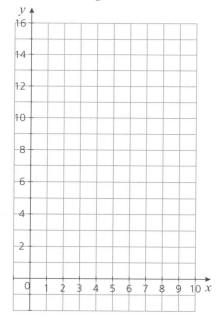

[6]

c State the possible solution(s) for the number of male and female athletes in this Olympic team.

.... [3]

Photocopying prohibited   *Cambridge IGCSE™ Core and Extended Mathematics Workbook*

15 Sequences

Exercise 15.1

1 For each arithmetic sequence:

    i deduce the common difference d
    ii write the term-to-term rule in its general form using the notation u_{n+1} and u_n
    iii give the formula for the nth term
    iv calculate the 50th term.

a 7, 11, 15, 19

    i .... [1]

    ii .... [1]

    iii .... [1]

    iv .... [1]

b 7, 4, 1, –2

    i .... [1]

    ii .... [1]

    iii .... [1]

    iv .... [1]

c –4, –3.5, –3, –2.5

    i .... [1]

    ii .... [1]

    iii .... [1]

    iv .... [1]

Exercise 15.2–15.3

Using a table if necessary, find the formula for the nth term for each sequence.

1 0, 7, 26, 63, 124

.... [2]

2 3, 10, 29, 66, 127

.... [3]

Exercise 15.4*

In questions 1–3 give the next two terms of the sequence.

1 $64, 32, 16, 8$ **2** $5000, 500, 50, 5$ **3** $10^3, 10^2, 10$

.................................[1]   ....[1]   .... [2]

4 The nth term of an exponential sequence is given by the formula $u_n = 2 \times 3^{n-1}$.

    a Calculate u_1, u_2 and u_3.

    .... [3]

    b What is the value of n if $u_n = 1458$?

    .... [2]

5 Part of an exponential sequence is given below:

$..., ..., 4, ..., ..., \frac{1}{16}$ where $u_3 = 4$ and $u_6 = \frac{1}{16}$

Calculate:

    a the common ratio r

    .... [2]

    b the value of u_1

    .... [2]

    c the formula for the nth term

    .... [2]

    d the 10th term, giving your answer as a fraction.

    .... [2]

Exercise 15.5*

In each question,

    a write down the rule for the nth term of the sequence by inspection
    b write down the next two terms of the sequence.

1 2, 6, 12, 20, 30

    a ....[3]   **b** .... [2]

2 $\frac{1}{2}$, 4, $13\frac{1}{2}$, 32, $62\frac{1}{2}$

    a .... [2]   **b** .... [2]

3 0, 2, 6, 14, 30

    a .... [2]   **b** .... [2]

4 3, 8, 17, 32, 57

    a ....[2]   **b** .... [2]

 Photocopying prohibited

16 Proportion*

Exercise 16.1*

1 a If d is proportional to p and the constant of proportionality is k, write an equation for d in terms of p.

.... [1]

b If $d = 10$ when $p = 5$ find k. .... [1]

c Find d when $p = 20$. .... [1]

d Find p when $d = 2$. .... [1]

2 a is inversely proportional to b.

a If k is the constant of proportionality, write an equation for a in terms of b.

.... [1]

b If $k = 20$, find a when $b = 40$. .... [1]

3 p is inversely proportional to q squared. If $q = 0.5$ when $p = 2$:

a Write an equation for p in terms of q. .... [1]

b Find p when $q = 5$. .... [1]

c If $p = 0.005$ find two values for q.

.... [2]

4 q is proportional to p squared and q is inversely proportional to r cubed. Using k as the final constant of proportionality, write an equation for p in terms of r.

....

.... [3]

Exercise 16.2*

1 a is proportional to the cube of b. If $b = 2$ when $a = 32$, find a when $b = 5$.

....

.... [2]

Exercise 16.3*

1 The power of an engine is proportional to the square of its mass. If an engine weighing 10 kg gives 200 bhp, find:

a the power of an engine weighing 30 kg

.... [2]

b the mass of an engine giving 5000 bhp.

.... [2]

2 The speed (v) in metres/second of a dam outlet is measured. It is proportional to the square root of the level (l) in metres indicated on a gauge. If $l = 64$ when $v = 24$, calculate v when $l = 12\,100$.

.... [3]

3 The force (f) newtons between two objects is inversely proportional to the square of the distance (l) metres between them. Two magnets attract with a force of 18 newtons when they are 2 cm apart. What is the force of attraction when they are 6 cm apart?

.... [3]

17 Graphs in practical situations

Exercise 17.2

1 Find the average speed of an object moving 60 m in 12 s.

....[1]

2 How far will an object travel during 2 h 18 min at 15 m/s?

....[2]

3 How long will an object take to travel 4.8 km at 48 m/s?

....[2]

Exercises 17.3–17.4

1 Two people, A and B, set off from points 300 m apart and travel towards each other along a straight road. Their movement is shown on the graph below:

a Calculate the speed of person A.

.... [1]

b Calculate the speed of person B when she is moving.

.... [1]

c Use the graph to estimate how far apart they are 50 seconds after person A has set off.

....[2]

d Explain the motion of person B in the first 20 seconds.

....[1]

e Calculate the average speed of person B during the first 60 seconds.

....[2]

2 A cyclist sets off at 09 00 one morning and does the following:

- Stage 1: Cycles for 30 minutes at a speed of 20 km/h.
- Stage 2: Rests for 15 minutes.
- Stage 3: Cycles again at a speed of 30 km/h for 30 minutes.
- Stage 4: Rests for another 15 minutes.
- Stage 5: Realises his front wheel has a puncture so walks with the bicycle for 30 minutes at a speed of 5 km/h to his destination.

a At what time does the cyclist reach his destination?

....[1]

b How far does he travel during stage 1?

....[1]

c Draw a distance–time graph on the axes below to show the cyclist's movement. Label all five stages clearly on the graph.

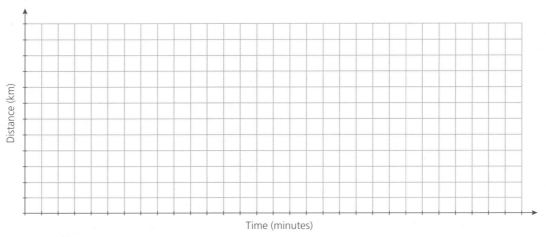

[5]

d Calculate the cyclist's average speed for the whole journey. Answer in km/h.

....[1]

Exercises 17.5–17.6*

1 Using the graphs, calculate the acceleration/deceleration in each case.

a

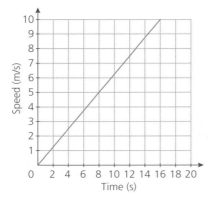

.... [1]

b

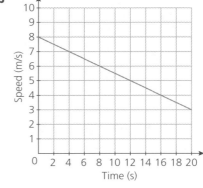

.... [2]

2 A sprinter is in training. Below is a graph of one of his sprints:

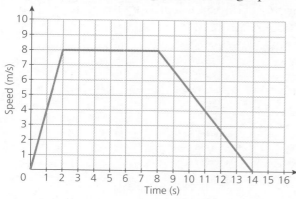

a Describe in words the sprinter's motion between the second and eighth second.

....

....

....[1]

b Calculate the acceleration/deceleration during the first two seconds.

.... [1]

c Calculate the acceleration/deceleration during the last phase of the sprint.

.... [1]

Exercise 17.7–17.8*

1 A stone is dropped off the top of a cliff. It accelerates at a constant rate of $10\,\text{m/s}^2$ for 4 seconds before hitting the water at the bottom of the cliff.

a Complete the table.

Time (s)	0	1	2	3	4
Speed (m/s)	0				

[2]

b Plot a speed–time graph for the 4 seconds it takes for the stone to drop.

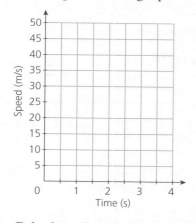

[2]

c Calculate the distance fallen by the stone between the first and third seconds.

.... [2]

d Calculate the height of the cliff.

.... [1]

2 Two objects, A and B, are at the same point when the time $t = 0$ s. At that point, object A accelerates from rest at a constant rate of 2 m/s^2 for 6 seconds. Object B is travelling at 15 m/s but decelerates at a constant rate of 3 m/s^2 until it comes to rest.

a On the axes, plot a speed–time graph for both objects.

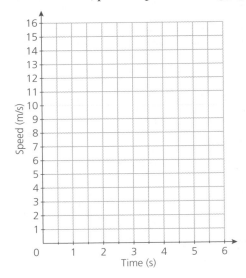

[4]

b After how many seconds are the two objects travelling at the same speed?

.... [1]

c Assuming both objects are travelling in the same straight line, calculate how far apart they are after 4 seconds.

.... [3]

3 A firework is launched vertically from the ground. The distance d (m) from the ground t (s) after launch is given by the formula $d = 80t - 5t^2$.

a Complete the table for the first 16 seconds of the firework's flight.

Time (s)	0	2	4	6	8	10	12	14	16
Distance from ground (m)	0				320				

[3]

b Plot the results on the graph.

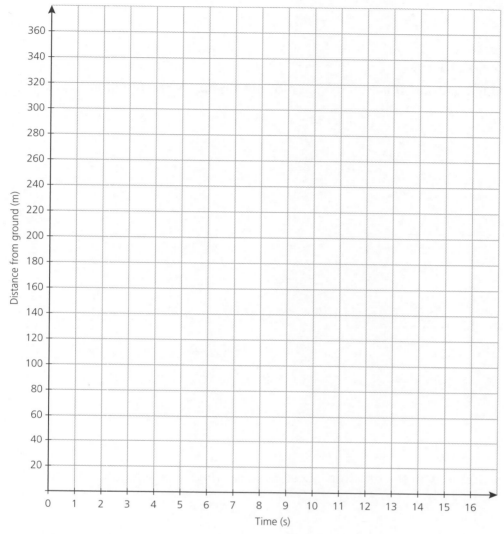

[2]

c From the graph, estimate the speed of the firework:

i 2 seconds after the launch

ii 8 seconds after launch.

.... [3]   .... [1]

18 Graphs of functions

Exercises 18.2–18.3

1 Solve the following quadratic function by first plotting a graph of the function.

$$2x^2 - 8x - 10 = 0$$

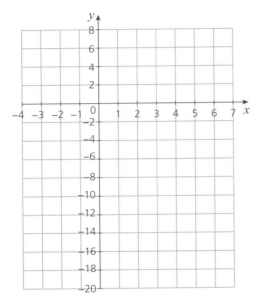

[3]

$x =$ .... [2]

2 Using the graph you drew in the previous question, solve the following quadratic equation. Show your method clearly.

$$2x^2 - 8x + 6 = 0$$

$x =$ .... [2]

Exercise 18.4*

1 a Plot the graph of the function $y = -(x-3)^2 + 5$ for $0 \leqslant x \leqslant 7$

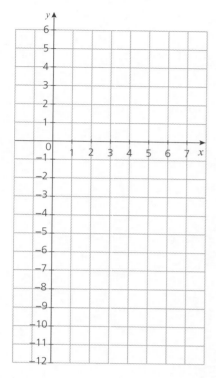

[3]

b State the coordinates of the turning point. ....[1]

2 For each of the functions a–c below, without plotting the graph, state:

   i whether the turning point is a maximum or a minimum

   ii the coordinates of the turning point.

a $y = (x-6)^2 - 4$ **i** ....[1]

                                                 ii ....[1]

b $y = x^2 - 3$ **i** ....[1]

                                                 ii ....[1]

c $y = -(x+8)^2 - 2$ **i** ....[1]

                                                 ii ....[1]

Exercise 18.5

1 $y = \dfrac{3}{2x}, x \neq 0$

a Complete the table

x	−4	−3	−2	−1	−0.5		0.5	1	2	3	4
y											

b On the axes blow, draw the graph of $y = \dfrac{3}{2x}$ for $-4 \leqslant x \leqslant -0.5$ and $0.5 \leqslant x \leqslant 4$. [2]

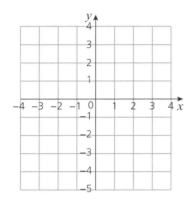

[2]

Exercises 18.6–18.7*

For each of the functions given:

a complete a table of values for x and $f(x)$

b plot a graph of the function.

1 $f(x) = \dfrac{1}{x^2} - x; \quad -4 \leqslant x \leqslant 3$

a

x	−4	−3	−2	−1	−0.5		0.5	1	2	3
y										

b

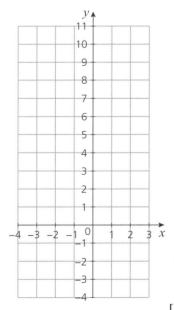

[2] [3]

2 $f(x) = 3^x - x - 2; \ -5 \le x \le 2$

a

x	−5	−4	−3	−2	−1	0	1	2
y								

b

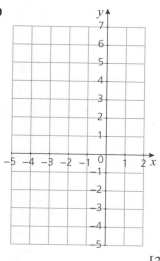

[2]

[3]

Exercise 18.8*

For each of the functions below:

   a plot a graph

   b calculate the gradient of the function at the point given.

1 $y = x^2 - x - 2; \ -2 \le x \le 3$

a

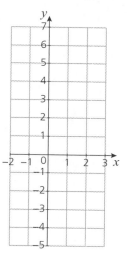

b Gradient where $x = 2$

[2]

.... [3]

2 $y = 2x^{-1} + x; \ 1 \le x \le 6$

a

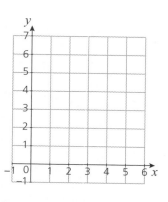

b Gradient where $x = 2$

[2]

.... [3]

Exercise 18.9*

1 a Plot the function $y = \dfrac{3}{x^2} - 2x$ for $-5 \leqslant x \leqslant 2$.

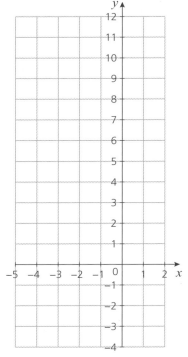

[3]

b Showing your method clearly, use the graph to solve the equation $x^3 + 4x^2 - 3 = 0$.

....[4]

2 a Plot the function $y = 3^x + \dfrac{1}{2}x$ for $-4 \leqslant x \leqslant 2$.

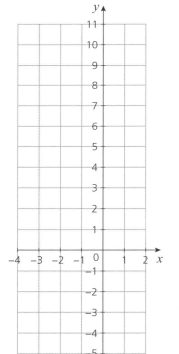

[3]

b Showing your method clearly, use the graph to solve the equation $2(3^x) - 3x - 8 = 0$.

....[4]

Exercise 18.10–18.11*

1 Sketch both linear functions below on the same axes. Label each one clearly, identifying where they cross each axis.

a $y = -\frac{1}{2}x + 5$

b $x - \frac{1}{2}y - 2 = 0$

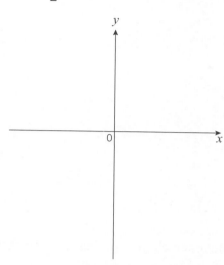

[4]

2 a Sketch both quadratic functions on the same axes below.

i $y = (x + 3)(2x + 2)$

ii $y = -x(x - 5)$

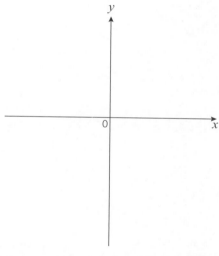

[4]

b Deduce the coordinates of the turning points.

i ....[2]   **ii** ....[2]

3 Sketch the cubic function $y = x^3 - 8x^2 + 16x$ on the axes below. Identify where it intersects with each axis.

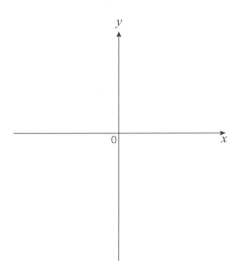

[4]

4 Sketch each of the functions below on the same axes. Label each clearly, indicating any key coordinates.

a $y = -\dfrac{1}{x}$

b $y = 3^x$

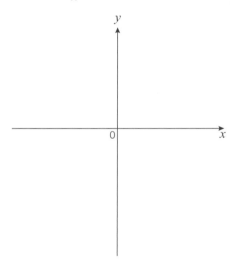

[4]

19 Differentiation and the gradient function*

Exercises 19.1–19.3*

1 a Find the gradient function of $f(x) = 2x^3$. .... [1]

b Calculate the gradient of this curve when:

   i $x = -2$ **ii** $x = 3$

   .... [1]   .... [1]

2 a Find the gradient function of $f(x) = \frac{1}{5}x^5$ .... [1]

b Calculate the gradient of this curve when:

   i $x = 0$ **ii** $x = 2$

   .... [1]   .... [1]

Exercises 19.4–19.5*

Differentiate each of the expressions in questions 1–3 with respect to x.

1 a $\frac{1}{2}x^3$ **b** $18x$ **c** $\frac{2}{5}$

.... [1]   .... [1]   .... [1]

2 a $-x^3 + 2x$ **b** $\frac{2}{3}x^3 - x + 1$ **c** $-\frac{5}{2}x^4 + \frac{1}{4}x^2 - 2$

.... [2]   .... [2]   .... [2]

3 a $\left(x - \frac{1}{2}\right)(2x + 1)$ **b** $\frac{3x^3 - x}{2x}$ **c** $\dfrac{\frac{1}{2}x^5 + \frac{1}{3}x^2}{\frac{3}{4}x^2}$

.... [3]   .... [3]   .... [4]

4 Calculate the derivative of each of the following functions:

a $v = (2 - r)^2$

.... [2]

b $p = q(4q+1)^2$

.... [3]

c $f = \dfrac{3g^2 + \frac{1}{2}g^3}{2g}$

.... [3]

Exercise 19.6*

Calculate the second derivative for each of the following:

1 $y = \frac{1}{2}x^3$

2 $y = 2x^5 - 3x^2 + 2$

.... [2]   .... [4]

3 $y = \frac{1}{2}(x^4 - x^5)x^{-2}$

4 $y = (2x^2 - 1)(x^2 + x)$

.... [3]   .... [4]

Exercise 19.7*

1 The population (P) of a colony of rabbits over a period of t weeks (where $t < 24$) is given by the formula $P = 6t^2 + 3t - \frac{1}{4}t^3 + 50$, where t is the time in weeks.

   a Calculate the rabbit population when:

      i $t = 0$

      .... [1]

      ii $t = 20$

      .... [2]

   b Calculate the rate of population growth $\frac{dP}{dt}$.

   .... [2]

   c Calculate the rate of population growth when:

      i $t = 5$ **ii** $t = 20$

      .... [1]   .... [1]

 Photocopying prohibited   *Cambridge IGCSE™ Core and Extended Mathematics Workbook*

d Complete the table for the rabbit population over the 22-week period.

t (weeks)	0	2	4	6	8	10	12	14	16	18	20	22
Population		78	142			430	518	582		590		358

[2]

e Plot a graph of the results on the axes.

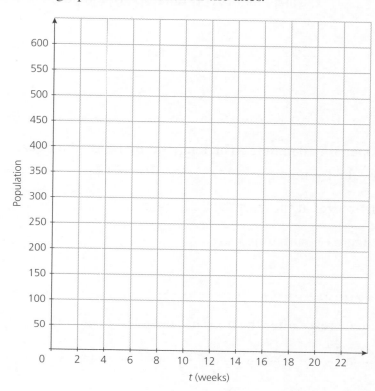

[3]

f With reference to the shape of your graph, explain your answers to part c.

....

.... [3]

Exercise 19.8*

1 A curve has equation $f(x) = \frac{1}{2}x^2 + x - 2$. Calculate the coordinate of the point P on the curve where the gradient is 8.

.... [3]

2 A curve has equation $f(x) = -\frac{2}{3}x^2 + 3x + 1$. Calculate the coordinate of the point Q on the curve where the gradient is –9.

....[3]

3 A curve has equation $f(x) = \frac{2}{3}x^3 + \frac{7}{2}x^2 + 2x$. Calculate the coordinate(s) of the point(s) on the curve where the gradient is 6.

....[5]

4 The door of a fridge freezer is left open for 10 minutes. The temperature inside the freezer ($T\,°C$) after time (t mins) is given by the formula $T = \frac{1}{40}t^3 - \frac{7}{8}t^2 + 10t - 18$.

a Calculate the temperature of the freezer at the start. .... [1]

b What is the rate of temperature increase with time?

....[2]

c Calculate the rate of temperature increase after the door has been left open for:

i 1 minute .... [1]

ii 10 minutes

....[2]

d Hence deduce the likely temperature of the room. Explain your answer.

....

....[2]

Exercise 19.9*

1 For the function $f(x) = \frac{1}{3}x^3 - 4x$:

a calculate the gradient function

....[2]

b calculate the equation of the tangent to the curve at the point $(3, -3)$.

.... [6]

2 A tangent T, drawn on the curve $f(x) = -2x^2 + 10x - 8$ at P, has an equation $y = -2x + 10$.

a Calculate the gradient function of the curve.

.... [2]

b What are the coordinates of the point P?

.... [5]

Exercise 19.10*

1 A curve has equation $f(x) = -\frac{1}{3}x^3 + \frac{5}{2}x^2 + 6x$.

a Calculate its gradient function.

.... [2]

b Calculate the coordinates of any stationary points.

.... [4]

c Determine the type of stationary points. Giving reasons for your answer(s).

....

.... [2]

20 Functions*

Exercise 20.1*

1 If f(x) = 3x + 3, calculate:

   a f(2) .... [1]   **b** f(4) .... [1]

   c f$\left(\frac{1}{2}\right)$ .... [1]   **d** f(–2) .... [1]

   e f(–6) .... [1]   **f** f$\left(-\frac{1}{2}\right)$ .... [1]

2 If f(x) = 2x – 5, calculate:

   a f(4) .... [1]   **b** f(7) .... [1]

   c f$\left(\frac{7}{2}\right)$ .... [1]   **d** f(–4.25) .... [1]

3 If g(x) = –x + 6, calculate:

   a g(0) .... [1]   **b** g(4.5) .... [1]

   c g(–6.5) .... [1]   **d** g(–2.3) .... [1]

Exercise 20.2*

1 If $f(x) = \frac{2x}{3} + 4$ calculate:

   a f(3)... [2]   **b** f(9)... [2]

   c f(–0.9) .... [2]   **d** f(–1.2) .... [2]

2 If g(x) = $\frac{7x}{2}$ – 3, calculate:

   a g(2) .... [2]   **b** g(0) .... [2]

   c g(–4) .... [2]   **d** g(–0.2) .... [2]

3 If h(x) = $\frac{-18x}{4}$ + 2, calculate:

   a h(1) .... [2]   **b** h(6) .... [2]

   c h(–4) .... [2]   **d** h(–0.8) .... [2]

Exercise 20.3*

1 If $f(x) = x^2 + 7$, calculate:

  a $f(11)$ .... [2]   **b** $f(1.1)$ .... [2]

  c $f(-13)$ .... [2]   **d** $f\left(\frac{1}{2}\right)$ .... [2]

  e $f\left(\sqrt{2}\right)$ .... [2]

2 If $f(x) = 2x^2 - 1$, calculate:

  a $f(5)$ .... [2]   **b** $f(-12)$ .... [2]

  c $f\left(\sqrt{3}\right)$ .... [2]   **d** $f\left(-\frac{1}{3}\right)$ .... [2]

3 If $g(x) = -5x^2 + 1$, calculate:

  a $g\left(\frac{1}{2}\right)$ .... [2]   **b** $g(-4)$ .... [2]

  c $g\left(\sqrt{5}\right)$ .... [2]   **d** $g\left(-\frac{3}{2}\right)$ .... [2]

Exercise 20.4*

1 If $f(x) = 3x + 1$, write down the following in their simplest form.

  a $f(x + 2)$ .... [3]

  b $f(2x - 1)$ .... [3]

  c $f(2x^2)$ .... [3]

  d $f\left(\frac{x}{2} + 2\right)$ .... [3]

2 If $g(x) = 2x^2 - 1$, write down the following in their simplest form.

  a $g(3x)$ .... [3]

  b $g\left(\frac{x}{4}\right)$ .... [3]

  c $g\left(\sqrt{2x}\right)$ .... [3]

  d $g(x - 5)$ .... [3]

Exercise 20.5*

Find the inverse of each of the following functions.

1 a $f(x) = x + 4$ .... [2]   **b** $f(x) = 5x$ .... [2]

2 a $g(x) = 3x - 5$ .... [3]   **b** $g(x) = \frac{5x}{2} - 1$ .... [3]

  c $g(x) = \dfrac{2(2x - 3)}{5}$

    ....

    .... [3]

Exercise 20.6*

1 If $f(x) = x - 1$, evaluate:

a $f^{-1}(2)$ .... [3]

b $f^{-1}(0)$ .... [1]

2 If $f(x) = 2x + 3$, evaluate:

a $f^{-1}(5)$ .... [3]

b $f^{-1}(-1)$ .... [1]

3 If $g(x) = 3(x - 2)$, evaluate $g^{-1}(12)$.

.... [3]

4 If $g(x) = \frac{x}{2} + 1$, evaluate $g^{-1}\left(\frac{1}{2}\right)$

.... [3]

Exercise 20.7*

1 Write a formula for $fg(x)$ in each of the following:

a $f(x) = 2x$, $g(x) = x + 4$ .... [3]

b $f(x) = x + 4$, $g(x) = x - 4$ .... [3]

2 Write a formula for $pq(x)$ in each of the following:

a $p(x) = 2x$, $q(x) = x + 1$ .... [3]

b $p(x) = x + 1$, $q(x) = 2x$ .... [3]

3 Write a formula for $jk(x)$ in each of the following:

a $j(x) = \frac{x - 2}{4}$, $k(x) = 2x$ .... [4]

b $j(x) = 6x + 2$, $k(x) = \frac{x - 3}{2}$ .... [4]

4 Evaluate $fg(2)$ in each of the following:

a $f(x) = 3x - 2$, $g(x) = \frac{x}{3} + 2$

.... [4]

b $f(x) = \frac{2}{x + 1}$, $g(x) = -x + 1$

.... [4]

21 Straight-line graphs

Exercises 21.6–21.8

1 Identify the coordinates of some of the points on the line and use these to find:

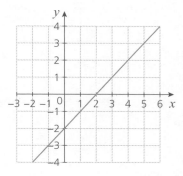

    a the gradient of the line

    .... [1]

    b the equation of the straight line

    .... [2]

Exercise 21.9

For each linear equation, calculate the gradient and *y*-intercept.

1 **a** $y = 4x - 2$ **b** $y = -(2x + 6)$

    ....

    .... [2]   .... [2]

2 **a** $y + \frac{1}{2}x = 3$ **b** $y - (4 - 3x) = 0$

    ....

    .... [3]   .... [3]

Exercise 21.10

1 Find the equation of the straight line parallel to $y = -2x + 6$ that passes through the point $(2, 5)$.

    ....

    .... [2]

Exercise 21.12

Solve the simultaneous equations:

    a by graphical means

    b by algebraic means.

1 $y = \frac{1}{2}x + 4$ and $y + x + 2 = 0$

    a   **b**

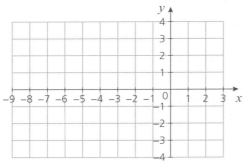

    .... [4]   .... [2]

2 $y + 3 = x$ and $3x + y - 1 = 0$

    a   **b**

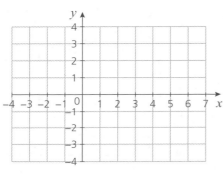

    .... [4]   .... [2]

Exercise 21.13*

In each of the following:

    a calculate the length of the line segment between each of the pairs of points to 1 d.p.

    b calculate the coordinates of the midpoint of the line segment.

1 (7, 3) and (7, 9)

    a   **b**

    .... [2]   .... [1]

 Photocopying prohibited   *Cambridge IGCSE™ Core and Extended Mathematics Workbook*

2 $(3, 5)$ and $(-2, 7)$

   a

                          b

.... [2]   .... [1]

3 $(-2, -4)$ and $(4, 0)$

   a

                          b

.... [2]   .... [1]

4 $\left(\frac{1}{2}, -3\right)$ and $\left(-\frac{1}{2}, 6\right)$

   a

                          b

.... [2]   .... [2]

Exercise 21.14*

Find the equation of the straight line which passes through each of the following pairs of points.

1 $(4, -2)$ and $(-6, -7)$ **2** $(0, 6)$ and $\left(\frac{1}{2}, 5\right)$

.... [2]   .... [3]

3 $(-2, 7)$ and $(3, 7)$ **4** $(-4, 2)$ and $\left(3, -\frac{3}{2}\right)$

.... [3]   .... [3]

5 $\left(\frac{1}{2}, -4\right)$ and $\left(\frac{1}{2}, -\frac{1}{2}\right)$ **6** $\left(2, -\frac{14}{3}\right)$ and $\left(4, -\frac{16}{3}\right)$

.... [3]   .... [3]

Exercise 21.15*

In each question, calculate:

    a the gradient of the line joining the points

    b the gradient of a line perpendicular to this line

    c the equation of the perpendicular line if it passes through the first point each time.

1 $(8, 3)$ and $(10, 7)$

    a .... [1]

    b .... [1]

    c ....

       ....

       .... [2]

2 $(3, 5)$ and $(4, 4)$

    a .... [1]

    b .... [1]

    c ....

       ....

       .... [2]

3 $(-3, -1)$ and $(-1, 4)$

   a .... [1]

   b .... [1]

   c ....
....
.... [2]

4 $(4, 8)$ and $(-2, 8)$

   a .... [1]

   b .... [1]

   c ....
....
.... [2]

5 $\left(\frac{1}{2}, \frac{5}{2}\right)$ and $\left(3, -\frac{5}{4}\right)$

   a .... [1]

   b .... [1]

   c ....
....
.... [2]

6 $\left(-\frac{7}{3}, \frac{1}{7}\right)$ and $\left(-\frac{7}{3}, \frac{3}{2}\right)$

   a .... [1]

   b .... [1]

   c ....
....
.... [2]

Photocopying prohibited

22 Geometrical vocabulary and construction

Exercise 22.2

1   Complete the table by entering either 'Yes' or 'No' in each cell.

	Rhombus	Parallelogram	Kite
Opposite sides equal in length.			
All sides equal in length.			
All angles right angles.			
Both pairs of opposite sides parallel.			
Diagonals equal in length.			
Diagonals intersect at right angles.			
All angles equal.			

[3]

Exercise 22.4

1   Using only a ruler and a pair of compasses, construct the following triangle XYZ.

XY = 5 cm, XZ = 3 cm and YZ = 7 cm.

[3]

Photocopying prohibited   *Cambridge IGCSE™ Core and Extended Mathematics Workbook*

23 Similarity and congruence

Exercise 23.1*

1

a Calculate the length a.

.... [2]

b Calculate the length b.

.... [3]

Exercise 23.2*

These five rectangles are each an enlargement of the previous one by a scale factor of 1.2.

1

a If the area of rectangle D is $100\,cm^2$, calculate to 1 d.p. the area of:

  i rectangle E

.... [1]

  ii rectangle A.

.... [2]

b If the rectangles were to continue in this sequence, which letter of rectangle would be the last to have an area below $500\,cm^2$? Show your method clearly.

.... [3]

2 A triangle has an area of $50\,cm^2$. If the lengths of its sides are all reduced by a scale factor of 30%, calculate the area of the reduced triangle.

.... [3]

Exercises 23.3–23.4*

1 A cube has a side length of $4.5\,cm$.

   a Calculate its total surface area.

   .... [2]

   b The cube is enlarged and has a total surface area of $1093.5\,cm^2$. Calculate the scale factor of enlargement.

   .... [3]

   c Calculate the volume of the enlarged cube.

   .... [2]

2 The two cylinders shown are similar.

   a Calculate the volume factor of enlargement.

   .... [1]

   b Calculate the scale factor of enlargement. Give your answer to 2 d.p.

   .... [2]

   c Calculate the value of x.

   .... [1]

3

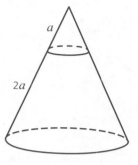

A large cone has its top sliced as shown in the diagram. The smaller cone is mathematically similar to the original cone.

a What is the scale factor of enlargement from the small cone to the original cone?

....

.... [1]

b If the original cone has a volume of 1350 cm³, calculate the volume of the smaller cone.

.... [3]

4 An architect's drawing has a scale of 1 : 50. The area of a garden on the drawing is 620 cm². Calculate the area of the real garden. Answer in m².

....

.... [4]

Exercise 23.5*

1 Quadrilaterals A and B are congruent.

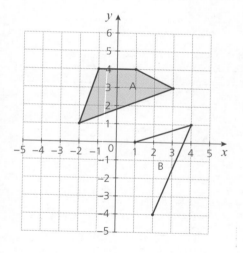

a Complete the diagram of shape B. [1]

b Give the coordinates of the missing vertex of shape B.

(........ , ...) [1]

Exercise 23.6*

1 All three angles and sides of triangle T are shown below.

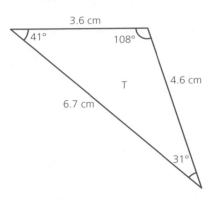

Explain, giving reasons, whether the triangles below are definitely congruent to triangle T.

a

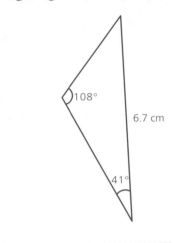

....

....

.... [2]

b

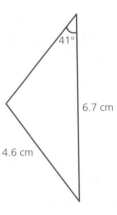

....

....

.... [2]

c

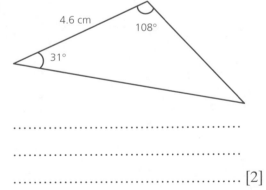

....

....

.... [2]

24 Symmetry*

Exercise 24.1*

1 On each pair of diagrams, draw a different plane of symmetry.

a A cuboid with a square cross-section.

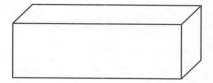 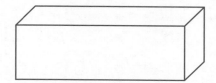

[2]

b A triangular prism with an equilateral triangular cross-section.

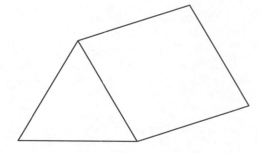 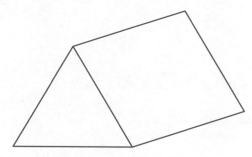

[2]

2 Determine the order of rotational symmetry of the cube, about the axis given.

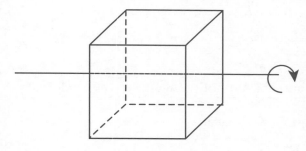

.... [2]

Exercise 24.2*

1 In the circle, O is the centre, AB = CD and X and Y are the midpoints of AB and CD respectively. Angle OCD = 50° and angle AOD = 30°.

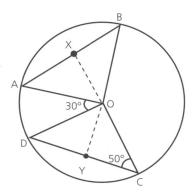

a Explain why triangles AOB and COD are congruent.

....

.... [2]

b What type of triangle is triangle AOB? .... [1]

c Calculate the obtuse angle XOY.

.... [3]

Exercise 24.3*

1 The diagram shows a circle with centre at O. XZ and YZ are both tangents to the circle.

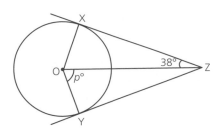

Calculate, giving detailed reasons, the size of the angle marked *p*.

.... [3]

25 Angle properties

Exercises 25.1–25.3

1 Calculate the size of each labelled angle.

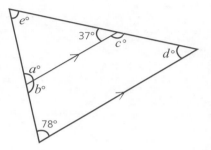

a = ... [1]

b = ... [1]

c = ... [1]

d = ... [1]

e = ... [1]

Exercise 25.5

1 In the diagram below, O marks the centre of the circle. Calculate the value of x.

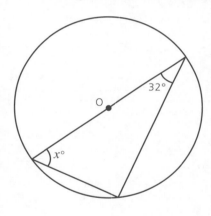

.... [2]

Exercise 25.7*

1 The pentagon below has angles as shown.

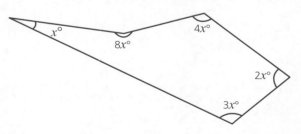

a State the sum of the interior angles of a pentagon.

.... [1]

b Calculate the value of *x*.

.... [2]

c Calculate the size of each of the angles of the pentagon.

.... [2]

2 The diagram shows an octagon.

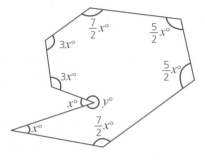

a Write the angle *y* in terms of *x*. .... [1]

b Write an equation for the sum of the interior angles of the octagon in terms of *x*.

.... [2]

c Calculate the value of *x*.

.... [2]

d Calculate the size of the angle labelled *y*.

.... [1]

Exercise 25.8*

In each diagram, O marks the centre of the circle. Calculate the value of the marked angles in each case.

1

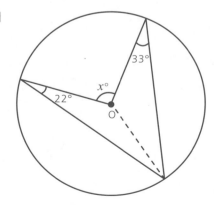

2

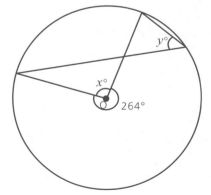

.... [3]   .... [3]

Cambridge IGCSE™ Core and Extended Mathematics Workbook

Exercise 25.9*

In the following, calculate the size of the marked angles.

1

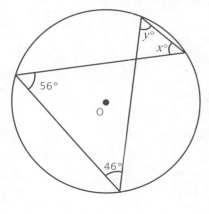

.... [3]

2

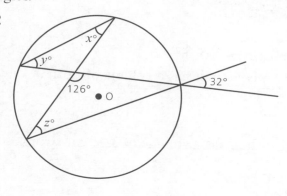

.... [3]

Exercise 25.10*

In the following, calculate the size of the marked angles.

1

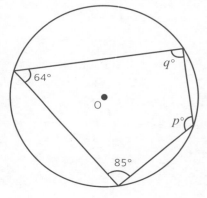

.... [2]

2

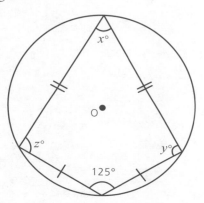

.... [2]

3

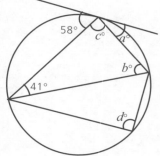

a = ... [1]

b = ... [1]

c = ... [1]

d = ... [1]

26 Measures

Exercises 26.1–26.5

1 **a** A container has a volume of $3.6\,m^3$. Convert the volume into cm^3.

.... [2]

b A box has a volume of $3250\,cm^3$. Convert the volume into:

i mm^3

.... [2]

ii m^3

.... [2]

27 Perimeter, area and volume

Exercises 27.1–27.5

1 A trapezium and parallelogram are joined as shown.

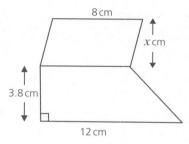

If the total area is 53.2 cm², calculate the value of x.

.... [3]

Exercises 27.6–27.9

1 A metal hand weight is made from two cubes and a cylinder joined as shown.

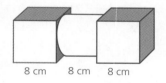

8 cm   8 cm   8 cm

Calculate the total volume of the shape.

.... [3]

Exercises 27.14–27.15

1 A hemispherical bowl, with an outer radius of 20 cm, is shown below. A sphere is placed inside the bowl. The size of the sphere is such that it just fits the inside of the bowl.

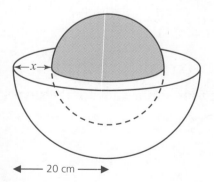

a Explain why the expression for the volume of the hemispherical bowl in terms of x can be written as $\frac{2}{3}\pi \times 20^3 - \frac{2}{3}\pi(20-x)^3$.

.... [2]

b Write an expression for the volume of the sphere in terms of x.

.... [2]

c If both the bowl and sphere have the same volume, show that $3(20-x)^3 = 8000$.

....

.... [2]

d Calculate the thickness x of the bowl.

.... [2]

Exercises 27.17–27.19

1 Two square-based pyramids are joined at their bases. The bases have an edge length of 6 cm.

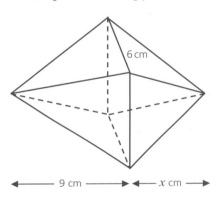

a Calculate the volume of the pyramid on the left.

.... [2]

b If the volume of the pyramid on the left is twice that of the pyramid on the right, calculate the value of x.

.... [2]

c By using Pythagoras as part of the calculation, calculate the total surface area of the shape.

.... [4]

Exercises 27.20–27.23

1 A cone has a base diameter of 8 cm and a sloping face length of 5 cm.

    a Calculate its perpendicular height.

    .... [2]

    b Calculate the volume of the cone.

    .... [2]

    c Calculate the total surface area of the cone.

    .... [4]

2 The two sectors shown are similar.

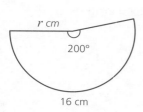 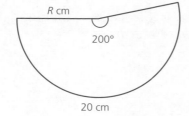

    a Calculate the length of the radius r.

    .... [2]

    b What is the value of R?

    .... [1]

Each sector is assembled to form two cones.

    c Calculate the volume of the smaller cone.

    .... [4]

    d Calculate the curved surface area of the large cone.

    .... [2]

3 A cone of base radius 10 cm and a vertical height of 20 cm has a cone of base radius 10 cm and a vertical height 10 cm removed from its inside as shown.

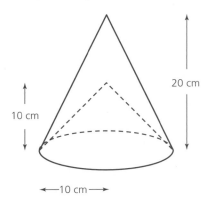

a Calculate the volume, in terms of π, of the small cone removed from the inside.

.... [2]

b Calculate the volume of the shape that is left (i.e. the volume of the large cone with the small cone removed). Give your answer in terms of π.

....

.... [2]

c Calculate the total curved surface area of the final shape.

.... [5]

28 Bearings

Exercise 28.1

1 A boat sets off from a point A on a bearing of 130° for 4 km to a point B. At B, it changes direction and sails on a bearing of 240° to a point C, 7 km away. At point C, it changes direction again and heads back to point A.

a Using a scale of 1 cm : 1 km, draw a scale diagram of the boat's journey.

[4]

b From your diagram work out:

i the distance AC .... [1]

ii the bearing of A from C .... [2]

29 Trigonometry

Exercises 29.1–29.3

Calculate the value of x in each diagram. Give your answers to 1 d.p.

1

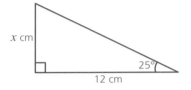

.... [2]

2

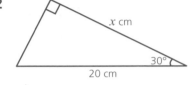

.... [2]

Exercises 29.4–29.5

1 Three towns, A, B and C, are positioned relative to each other as follows:

- Town B is 68 km from A on a bearing of 225°.
- Town C is on a bearing of 135° from A.
- Town C is on a bearing of 090° from B.

a Drawing a sketch if necessary, deduce the distance from A to C.

[2]

b Calculate the distance from B to C.

....

.... [2]

Exercise 29.6*

1 A point A is at the top of a vertical cliff, 25 m above sea level. Two points X and Y are in the sea. The angle of elevation from Y to A is 23°. Y is twice as far from the cliff as X.

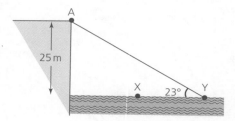

a Calculate the horizontal distance of Y from the foot of the cliff.

.... [2]

b Calculate the angle of depression from A to X to the nearest whole number.

.... [3]

c Calculate the ratio of the distances AX : AY. Give your answer in the form 1 : *n* where *n* is given to 1 d.p.

.... [5]

2 A tall vertical mast is supported by two wires, AC and BC. Points A and B are 2.5 m and 6 m above horizontal ground level respectively. Horizontally, the mast is 20 m and 27 m from A and B respectively. The angle of elevation of C from A is 30°.

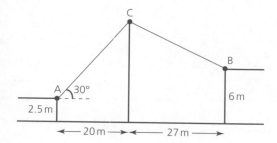

a Calculate the height of the mast.

.... [3]

b Calculate the angle of elevation of C from B.

.... [3]

c Calculate the shortest distance between A and B.

.... [2]

Exercises 29.7–29.8*

1 Express the following in terms of the sine of another angle between 0° and 180°:

a sin 86°.. [1]   **b** sin 158°... [1]

2 Express the following in terms of the cosine of another angle between 0° and 180°:

a cos 38°.. [1]   **b** cos 138°... [1]

3 Find the two angles between 0° and 180° which have the following sine. Give each answer to the nearest degree.

a 0.37.. [2]   **b** 0.85.. [2]

4 The cosine of which acute angle has the same value as:

a −cos 162°... [2]   **b** −cos 136°.. [2]

Exercise 29.9*

1 Solve the following equations, giving all the solutions for θ in the range $0 \leqslant \theta \leqslant 360°$

a $\sin \theta = -\frac{\sqrt{3}}{2}$

.... [2]

b $\tan \theta = -\sqrt{3}$

.... [2]

2 Calculate the value of $\cos \theta$ in the diagram below. Leave your answer in surd form.

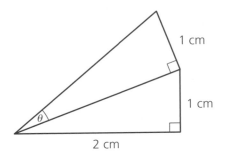

.... [5]

30 Further trigonometry*

Exercises 30.1–30.2*

1 Calculate the length of the side marked x.

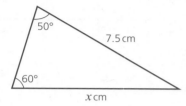

.... [2]

2 Calculate the length of the side marked x.

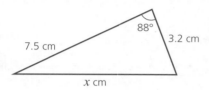

.... [2]

3 Calculate the length of the side marked x.

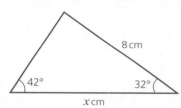

.... [2]

4 Calculate the size of the angle marked θ.

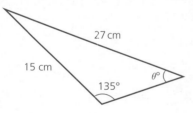

.... [2]

5 Calculate the size of the angle θ below, given that it is an obtuse angle (between 90° and 180°).

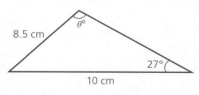

.... [4]

Exercise 30.3*

1 A bird, B, flies above horizontal ground. The bird is 126 m from point A on the ground and 88 m from a point C also on the ground. Given that the distance between A and C is 100 m, calculate:

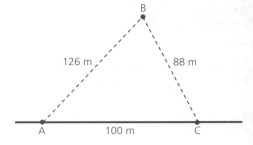

a the angle of elevation from A to B

.... [3]

b the height of the bird above the ground.

.... [2]

Exercise 30.4*

1 Calculate the area of the triangle.

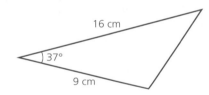

.... [2]

2 A triangle and rectangle are joined as shown below. If the total area of the combined shape is 110 cm², calculate the length of the side marked x.

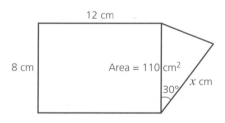

.... [4]

3 ABC is a triangle. AB = 15 cm, CAB = 62° and ACB = 82°.

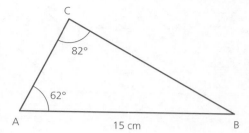

Calculate the vertical height of C above the base AB. Give your answer to 1 d.p.

.... [5]

Exercises 30.5–30.6*

1 The cone has its apex P directly above the centre of the circular base X. PQ = 12 cm and angle PQX = 72°.

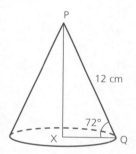

a Calculate the height of the cone.

.... [2]

b Calculate the circumference of the base.

.... [3]

2 ABCDEFGH is a cuboid. AD = 8 cm, DH = 5 cm and X is the midpoint of CG. Calculate:

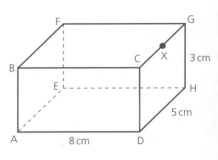

a the length EG

.... [2]

b the angle EGA

.... [2]

c the length AX

.... [2]

d the angle AXE.

.... [3]

3 ABCDEF is a right-angled triangular prism. AB = 3 cm, AC = 4 cm, BE = 9 cm and point X divides BE in the ratio 1 : 2.

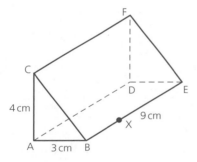

Calculate:

a the length BC

.... [1]

b the angle BXC

.... [2]

c the length XF

.... [2]

Photocopying prohibited
Cambridge IGCSE™ Core and Extended Mathematics Workbook

d the angle between XF and the plane ABDE.

.... [3]

4 The diagram below shows a right pyramid, where E is vertically above X. AB = 5 cm, BC = 4 cm, EX = 7 cm and P is the midpoint of CE.

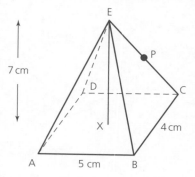

Calculate:

a the length AX

.... [2]

b the angle XCE

.... [2]

c the length XP.

.... [5]

31 Vectors

Exercises 31.2–31.3

In questions 1 and 2 consider the following vectors:

$$\mathbf{a} = \begin{pmatrix} 2 \\ 0 \end{pmatrix} \qquad \mathbf{b} = \begin{pmatrix} -3 \\ 1 \end{pmatrix} \qquad \mathbf{c} = \begin{pmatrix} 3 \\ -2 \end{pmatrix}$$

1 Express the following as a single column vector:

    a $3\mathbf{a}$ .... [2]

    b $2\mathbf{c} - \mathbf{b}$ .... [2]

    c $\frac{1}{2}(\mathbf{a} - \mathbf{b})$ .... [2]

    d $-2\mathbf{b}$ .... [2]

2 Draw vector diagrams to represent the following.

    a $2\mathbf{a} + \mathbf{b}$ **b** $-\mathbf{c} + \mathbf{b}$

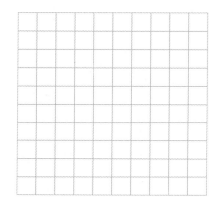
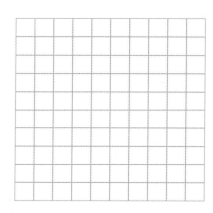

                                                                                [3]   [3]

Exercise 31.4*

Consider the vectors:

$$\mathbf{a} = \begin{pmatrix} -2 \\ 0 \end{pmatrix} \qquad \mathbf{b} = \begin{pmatrix} -3 \\ 2 \end{pmatrix} \qquad \mathbf{c} = \begin{pmatrix} 4 \\ -4 \end{pmatrix}$$

1 Calculate the magnitude of the following, giving your answers to 1 d.p.

    a $\mathbf{a} + \mathbf{b} + \mathbf{c}$

    .... [3]

 Photocopying prohibited   *Cambridge IGCSE™ Core and Extended Mathematics Workbook*

b $2\mathbf{b} - \mathbf{c}$

.... [3]

Exercises 31.5–31.7*

1 Consider the vector diagram shown. If $\overrightarrow{AB} = \mathbf{a}$ and $\overrightarrow{BC} = \mathbf{b}$ express the following in terms of **a** and/or **b**.

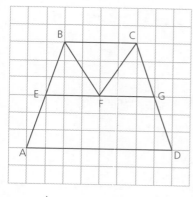

a $\overrightarrow{AE}$ .... [1]

b $\overrightarrow{EG}$ .... [1]

c $\overrightarrow{AF}$

.... [2]

d $\overrightarrow{CG}$

.... [3]

2 In the diagram, $\overrightarrow{AB} = \mathbf{a}$, $\overrightarrow{BD} = \mathbf{b}$, D divides the line BC in the ratio 2 : 1 and E is the midpoint of AC. Express the following in terms of **a** and **b**:

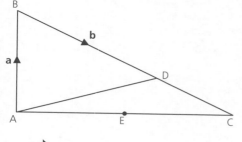

a $\overrightarrow{AD}$ .... [1]

b $\overrightarrow{DC}$ .... [1]

c $\overrightarrow{AC}$

.... [2]

d $\overrightarrow{ED}$

.... [2]

32 Transformations

Exercises 32.3–32.4

1 In the following, the object and centre of rotation are given. Draw the object's image under the stated rotation.

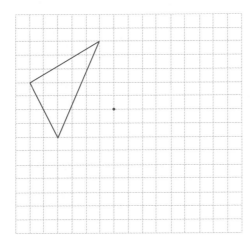

Rotation 180°   [2]

2 In the following, the object (unshaded) and image (shaded) have been drawn. Mark the centre of rotation and calculate the angle and direction of rotation.

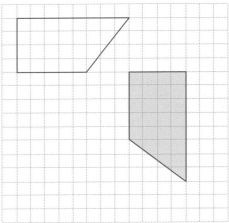

[2]

Exercise 32.5–32.6

1 In the diagram, object A has been translated to each of the images B, C and D. Give the translation vector in each case.

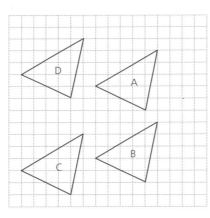

B = ... [1]

C = ... [1]

D = ... [1]

Photocopying prohibited   *Cambridge IGCSE™ Core and Extended Mathematics Workbook*

Exercise 32.10*

1 In this question, draw each transformation on the same grid and label the images clearly.

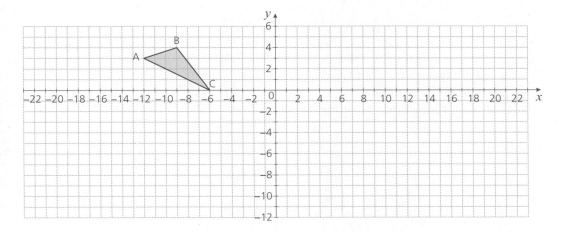

a Map the triangle ABC onto $A^1B^1C^1$ by an enlargement scale factor –2, with the centre of the enlargement at (–4, 2). [2]

b Map the triangle $A^1B^1C^1$ onto $A^2B^2C^2$ by a reflection in the line $x = 4$. [2]

c Map the triangle $A^2B^2C^2$ onto $A^3B^3C^3$ by a rotation if 180°, with the centre of rotation at (–4, 2). [2]

33 Probability

Exercises 33.1–33.5

1 In a class there are 23 girls and 17 boys. They enter the room in a random order. Calculate the probability that the first student to enter will be:

a a girl

.... [2]

b a boy

.... [2]

2 Two friends are standing in a hall with many other people. A person is picked randomly from the hall. How many people are in the hall if the probability of either of the friends being picked is 0.008?

.... [2]

Exercise 34.1

1 A fair 6-sided dice with faces numbered 1–6 and a fair 4-sided dice with faces numbered 1–4 are rolled. Use a two-way table if necessary to find:

   a the probability that both dice show the same number

   .... [2]

   b the probability that the number on one dice is double the number on the other.

   .... [2]

Exercise 34.4*

1 A cinema draws up a table showing the age (A) and gender of people watching a particular film.

	$A < 10$	$10 \leqslant A < 18$	$A \geqslant 18$	Total
Male	45	60	55	160
Female	35	75	80	190
Total	80	135	135	350

A person is picked at random; calculate the probability that:

   a they are under 10 years old .... [1]

   b they are a male not under 10 years old .... [1]

   c they are $\geqslant$ 18 years old, given that they are female

   .... [2]

   d they are a male, given that they are aged $10 \leqslant A < 18$.

   .... [2]

2 A player in a football team analyses her games and realises that she takes a penalty kick (P) in 15% of matches. If she takes a penalty, she scores (S) in 90% of the matches. If she does not take a penalty, she only scores in 30% of the matches.

   a Complete the tree diagram.

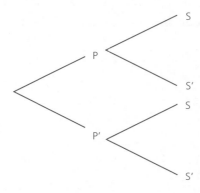

[2]

   b Calculate the probability that:

      i   she scores .... [2]

      ii she takes a penalty, given that she doesn't score.

          ....

          .... [3]

35 Mean, median, mode and range*

Exercise 35.3*

1 A school holds a sports day. The time taken for a group of students to finish the 1 km race is shown in the grouped frequency table.

Time (min)	4–	5–	6–	7–	8–9
Frequency	1	4	8	7	2

   a How many students completed the race? .... [1]

   b Estimate the mean time it took for the students to complete the race. Answer to the nearest second.

      .... [4]

36 Collecting and displaying data

Exercises 36.1–36.3

1 In 2012, the Olympics were held in London. 15 athletes were chosen at random and their height (cm) and mass (kg) were recorded. The results are:

Height (cm)	Mass (kg)
201	120
203	93
191	97
163	50
166	63
183	90
182	76
183	87

Height (cm)	Mass (kg)
166	65
160	41
189	82
198	106
204	142
179	88
154	53

a What type of correlation (if any) would you expect between a person's height and mass? Justify your answer.

.... [2]

b Plot a scatter graph on the grid below.

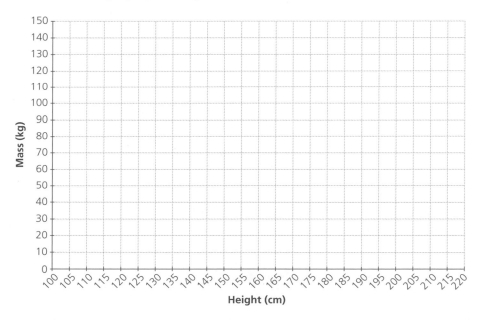

[3]

c i Calculate the mean height of the athletes.

.... [1]

ii Calculate the mean mass of the athletes.

.... [1]

 Photocopying prohibited *Cambridge IGCSE™ Core and Extended Mathematics Workbook*

iii Plot the point representing the mean height and mean mass of the athletes. Label it M.

[1]

d Draw a line of best fit for the data. Make sure it passes through M.

[1]

e i From the results you have plotted, describe the correlation between the height and mass of the athletes.

.... [1]

ii How does the correlation compare with your prediction in a?

.... [1]

Exercises 36.4–36.5*

1 The ages of 80 people, selected randomly, travelling on an aeroplane are given in the grouped frequency table:

Age (years)	0–	15–	25–	35–	40–	50–	60–	80–100
Frequency	10	10	10	10	10	10	10	10
Frequency density								

a Complete the table by calculating the frequency density.

[2]

b Represent the information as a histogram on the grid below.

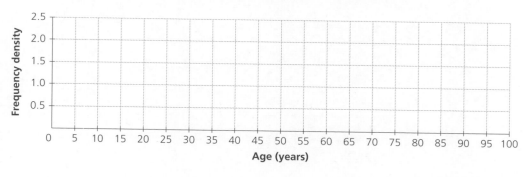

[3]

37 Cumulative frequency and box plots*

Exercises 37.1–37.2*

1 A candle manufacturer tests the consistency of their candles by randomly selecting 160 candles, lighting them and recording how long they last in minutes. The results are shown in the grouped frequency table:

Time (min)	140–	150–	160–	170–	180–	190–	200–	210–220
Frequency	5	20	45	30	25	20	10	5
Cumulative frequency								

a Complete the table by calculating the cumulative frequency. [1]

b Plot a cumulative frequency graph on the axes below.

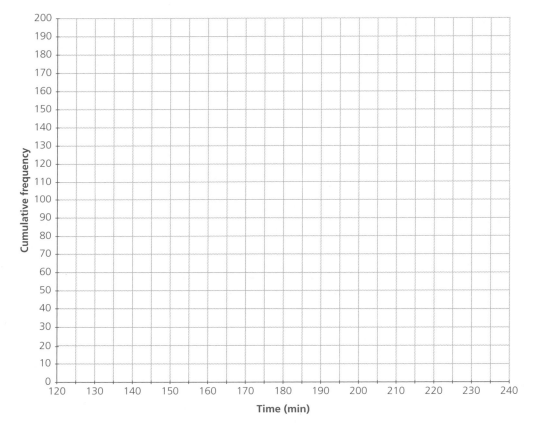

[3]

c From your graph, estimate the median amount of time that the candles last.

.... [1]

Cambridge IGCSE™ Core and Extended Mathematics Workbook

d From your graph estimate:

   i the upper quartile time .... [1]

   ii the lower quartile time .... [1]

   iii the interquartile range. .... [1]

e Draw a box-and-whisker plot for this data.

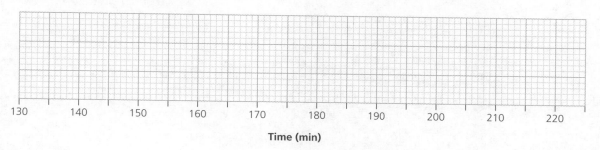

          Time (min)

[3]

f The manufacturer is aiming for the lifespans of the middle 50% of candles to not differ by more than 30 minutes. Explain, giving your justification, whether the data supports this aim.

....

.... [2]

Reinforce learning and deepen understanding of the key concepts covered in the revised syllabus; an ideal course companion or homework book for use throughout the course.

» Develop and strengthen skills and knowledge with a wealth of additional exercises that perfectly supplement the Student's Book.

» Build confidence with extra practice for each lesson to ensure that a topic is thoroughly understood before moving on.

» Ensure students know what to expect with hundreds of rigorous practice and exam-style questions.

» Keep track of students' work with ready-to-go write-in exercises.

» Save time with all answers available online in the Online Teacher's Guide.

Use with *Cambridge IGCSE™ Mathematics Core and Extended 4th edition*
9781510421684

For over 25 years we have been trusted by Cambridge schools around the world to provide quality support for teaching and learning. For this reason we have been selected by Cambridge Assessment International Education as an official publisher of endorsed material for their syllabuses.

This resource is endorsed by Cambridge Assessment International Education

✓ Provides learner support for the Cambridge IGCSE™ and IGCSE™ (9–1) Mathematics syllabuses (0580/0980) for examination from 2020

✓ Has passed Cambridge International's rigorous quality-assurance process

✓ Developed by subject experts

✓ For Cambridge schools worldwide

WORLD LAND TRUST™
www.carbonbalancedprint.com
CBP2250

ISBN 978-1-5104-2170-7

9 781510 421707

MIX
Paper from responsible sources
FSC™ C104740

HODDER EDUCATION
www.hoddereducation.com